200 Challenging Math Problems

Every 2nd Grader Should Know

This book belongs to:

_____

Grade:_____

200 Challenging Math Problems

every 2nd grader should know

New edition 2017
Copyright Learn 2 Think Pte. Ltd.

Published by:
Learn 2 Think Pte. Ltd.

ISBN: 978-981-07-2763-5

Master Grade 2 Math Problems

Introduction:

Solving math problems is core to understanding math concepts. When Math problems are presented as real-life problems students get a chance to apply their Math knowledge and concepts they have learnt. Word problems progressively develop a student's ability to visualize and logically interpret Mathematical situations.

This book provides numerous opportunities to students to practice their math skills and develop their confidence of being a lifelong problem solver. The multi-step problem solving exercises in the book involve several math concepts. Student will learn more from these problems solving exercises than doing ten worksheets on the same math concepts. The book is divided into 7 chapters. The last chapter of the book explains step wise solutions to all the problems to reinforce learning and better understanding.

How to use the book:

Here is a suggested plan that will help you to crack every problem in this book and outside.

Follow these 4 steps and all the Math problems will be a NO PROBLEM!

Read the problem carefully:

- ✎ What do I need to find out?
- ✎ What math operation is needed to solve the problem? For example addition, subtraction, multiplication, division etc.
- ✎ What clues and information do I have?
- ✎ What are the key words like sum, difference, product, perimeter, area, etc.?
- ✎ Which is the non-essential information?

Decide a plan

- ✎ Develop a plan based on the information that you have to solve the problem. Consider various strategies of problem solving:
- ✎ Drawing a model or picture
- ✎ Making a list
- ✎ Looking for pattern
- ✎ Working backwards
- ✎ Guessing and checking
- ✎ Using logical reasoning

Solve the problem:

Carry out the plan using the Math operation or formula you choose to find the answer.

Check your answer

- ✎ Check if the answer looks reasonable
- ✎ Work the problem again with the answer
- ✎ Remember the units of measure with the answer such as feet, inches, meter etc.

Master Grade 2 Math Problems

Note to the Teachers and Parents:

✎ Help students become great problem solvers by modelling a systematic approach to solve problems. Display the 'Four step plan of problem solving' for students to refer to while working independently or in groups.

✎ Emphasise on some key points:

✎ Enable students to enjoy the process of problem solving rather than being too focused on finding the answers.

✎ Provide opportunities to the students to think; explain and interpret the problem.

✎ Lead the student or the group to come up with the right strategy to solve the problem.

✎ Discuss the importance of showing steps of their work and checking their answers.

✎ Explore more than one possible solution to the problems.

✎ Give a chance to the students to present their work.

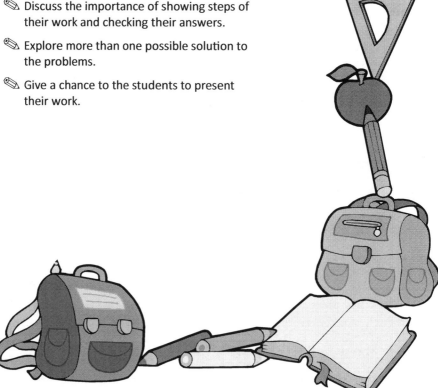

Contents

PROBLEM 1

I am thinking of a two digit number. The sum of its two digits is 12. The product of its two digits is 27. Write down all the possible numbers which I may be thinking of.

$1 + 11 = 12$
$2 + 10 = 12$

$\underline{3} \times \underline{9} = 27$

$3 + 9 = 12$
$4 + 8 = 12$
$5 + 7 = 12$
$6 + 6 = 12$

$1 \times 11 = 11$
$2 \times 10 = 20$
$3 \times 9 = \underline{27}$
$4 \times 8 = 24$
$5 \times 7 = 35$
$6 \times 6 = 36$

Answer: $3 \times 9 = 27$ so the asiwer is 3×9.

PROBLEM 2

Alison is reading a book. She noted that the sum of the 2 consecutive pages that she is reading now is 35. What are the page numbers of these two pages?

$15+16=31$
$16+17=33$
$17+18=35$
14

$17+18=35$

Answer: $17+18=35$ so the aswer is 35.

PROBLEM 3

On a farm there are a total of 7 goats and hens. They have a total of 24 legs. How many goats and hens are there on the farm?

2 hens , goats

7

$5 \times 2 = 10 + 2 \times 4 = 8 = 18$

$4 \times 2 = 8 + 3 \times 4 = 12 = 20$

$3 \times 2 = 6 + 4 + 4 = 16 = 22$

$2 \times 2 = 4 + 5 \times 4 = 20 = 24$

Answer: 2hens and 5 goats.

PROBLEM 4

Guess my number.

I am more than 25 but less than 40. My ones digit is twice my tens digit. My digits adds up to an odd number.

26 31
27 32
28 33
29 34
30 35
    (36)
    37
    38
    39

Answer: 36

PROBLEM 5

The sum of 3 odd numbers is 21.
Find the 3 numbers.

Write down all the odds number
under (21)

1 3 5 7 9 11 13 15 17 19

1 + 9 + 11 =

1 + 3 + 17 =

1 + 5 + 15 =

1 + 7 + 13 =

3 + 7 + 11 =

Answer: ……………………………

The difference between two number is <u>142</u>. If the bigger number is 196, what is the smaller number?

$$196 - 54 = 142$$

$$\begin{array}{r} 196 \\ -142 \\ \hline 54 \end{array}$$

Answer: ....54....................

PROBLEM 7

I am a mystery number greater than 6 but less than 16. I am an odd number. When you count by 5's, you say my name. What mystery number am I?

5 — 10 ⑮ — 16

7   15
8
9
10
11
12
13
14

Answer: ....15....................

[9875] 5 _(handwritten)_

PROBLEM 8

What is the largest 4-digit odd
number you can make that has
no repeating digits?

(handwritten work)

1234 10
6008
9,000

1, 3, 5, 7, 9

largest 4-
9875

9875

Answer: 9875

PROBLEM 9

Four children were lining up to buy lunch from school canteen. Their names are Lucy, Bill, Jim, and Mary. Lucy was first and Bill was last. Mary was behind Lucy but in front of Jim. Who was third in the line?

L, M, J, B

Answer: Jim

PROBLEM 10

10 people were in a queue to return their library books. Joanne was the 7th person from the front. Bob was the 6th person from the back. How many people were standing between Bob and Joanne?

$$10 - 7 + 1 = 4$$

Answer: ...4.........................

PROBLEM 11

20 students are lining up to board the bus. Jack is in 12th place from the front. Ann is at the 5th place from the back. How many people are standing between Jack and Ann?

Answer: 3

PROBLEM 12

20 books are placed on a stack. The 8th book from the top is a dictionary. The 16th book from the bottom is a phone directory. How many books are there in between the dictionary and the phone directory?

Answer: 2

PROBLEM 13

Charles is in the 5th place in a queue to buy movie tickets. Tony is last in the queue. There are 5 people in between Tony and Victor and 2 people in between Charles and Victor.

a) What position is Victor at in the queue?

b) How many people are there in the queue?

Answer: Victor is in 8 place. Their are 11 people.

PROBLEM 14

Doreen is in the 7th place in a queue to buy entrance tickets to a zoo. Xavier is in the last in the queue. There are 8 people between Xavier and Ronald and 5 people between Doreen and Ronald.

a) At what position is Ronald in the queue?

b) How many people are there in the queue?

Answer: Ronalds in 12th place their are 16 people in line.

PROBLEM 15

20 athletes are running one behind the other along a jogging track. The runner in the 15th position is Tom. The 12th runner from the back is Fred. How many runners are there in between Tom and Fred?

Fred tom   tom

Answer: 5

10 trees are planted along a straight road. The 8th tree from the front is a palm tree. The 5th tree from the back is a mango tree. How many trees are planted between the palm tree and the mango tree?

Answer: ……………………….

PROBLEM 17

There are 120 red marbles and 130 blue marbles in a box. Another 200 green marbles are added into the box. How many marbles are there in the box now?

Answer: ....

PROBLEM 18

Paul read 20 pages of a magazine on Sunday. He read 14 more pages on Monday than on Sunday. If there were 40 unread pages, how many pages were there in the magazine?

Answer: ....

PROBLEM 19

There were 196 pages in a book. Jason read 18 pages on Monday, 28 pages on Tuesday and 38 pages on Wednesday. If Jason continues to read in this pattern, how many days would he take to finish reading the entire book?

Answer: ....

PROBLEM 20

Janice has 120 mangoes and 240 tomatoes. Janice has 82 mangoes fewer than Karen. What is the total number of mangoes and tomatoes that Karen and Janice have?

Answer: ....

PROBLEM 21

Bryan sold 120 chicken on Monday and 200 chicken on Tuesday. He packed the rest of his chicken equally into 10 boxes of 10 chicken each. How many chicken did he have at first?

Answer: ....

PROBLEM 22

Florence used 16 bananas to bake some cakes and 14 bananas to bake some cookies. She had 62 bananas left. How many bananas did she have at first?

Answer: ....

PROBLEM 23

Kathy has 40 stamps. Sharon has 18 more stamps than Kathy. Joe has 35 more stamps than Sharon. How many stamps do Joe and Sharon have altogether?

Answer: ....

PROBLEM 24

A clock chimes once at 1 o'clock, twice at 2 o' clock and thrice at 3 o'clock. How many times would it chime by 7 o' clock?

Answer: …………………………

PROBLEM 25

What is the largest number that can be written using the following digits: 4, 8, 5? Use each digit only once in each number.

Answer: …………………………

There were 130 cars in a car park. 20 cars drove out of the car park and 42 cars drove in. How many cars were there in the car park in the end?

Answer: ....

PROBLEM 27

There were 120 people in a shopping center. 35 of them were children, 20 of them were women and the rest were men. How many more men than women were there in the shopping center?

Answer: ....

PROBLEM 28

Kevin had some stickers. His mother gave him 14 stickers. After he gave 18 stickers to Jerry, he had 70 stickers left. How many stickers did he have at first?

Answer: ....

PROBLEM 29

Sam and Alex had 20 marbles altogether. After giving four marbles to Alex, Sam and Alex had an equal number of marbles. How many marbles did Sam have in the beginning?

Answer: ....

PROBLEM 30

There were 120 passengers on a train. At the first stop, 14 passengers got on and 50 passengers got off. At the second stop, some passengers got off the train but none got on. There were then 24 passengers left on the train. What was the number of passengers who got off the train at the second stop?

Answer: ....

PROBLEM 31

There were 160 potatoes in a box. 18 rotten ones were thrown away and some were baked. There were 120 potatoes left in the box in the end. How many potatoes were baked?

Answer: ....

PROBLEM **32**

A box contains 40 cupcakes. There are 16 strawberry cupcakes, 22 blueberry cupcakes and some chocolate cupcakes. How many chocolate cupcakes are there in 6 such boxes?

Answer: …………………………

PROBLEM 33

Henry had 70 stickers. Jane had 95 stickers more than him and Patrick had 15 stickers fewer than Jane. How many stickers did they have altogether?

Answer: ....

Sam had some coins. He gave away 14 coins to Gerald. His father gave him 18 more coins. He finally has a total of 30 coins. How many coins did Sam have at first?

Answer: ....

Karen has 48 Australia stamps.
She has 14 fewer England
stamps than Australia stamps.
How many stamps does she have
altogether?

Answer: …………………………

Jennifer had 52 stamps. She gave 34 stamps to her brother and bought some new stamps. She had 74 stamps in the end. How many new stamps did Jennifer buy?

Answer: ………………………

PROBLEM 37

David had 130 pears. He threw away 85 rotten ones and bought another 72 pears. How many pears did he have in the end?

Answer: …………………………

PROBLEM 38

Mr. Ted bought 7 packs of pencils for his class. The total cost of pencils is $98. If every pack of pencils cost the same amount, how much does each pack cost?

Answer: …………………………

PROBLEM 39

Susan had 130 stamps. She gave some stamps to her brother and 26 stamps to her sister. She had 20 stamps left. How many stamps did she give to her brother?

Answer: ....

PROBLEM 40

There were 450 people at a party. There were 130 men, 120 women and the rest were children. How many children were there at the party?

Answer: ....

PROBLEM 41

Tim had 36 toy cars. Jason had 20 more toy cars than him and Kevin had 50 toy cars fewer than Jason. How many toy cars did they have altogether?

Answer: ....

PROBLEM 42

Mark had 80 bananas. Simon has 20 bananas lesser than Mark. Sarah had 15 more bananas than Simon. If Sarah sold 30 bananas, how many bananas were left with her?

Answer: ………………………

PROBLEM 43

Jessica, Mary and Thomas had 150 chocolates altogether. Jessica had 60 chocolates. Mary had 10 chocolates more than Jessica. How many chocolates did Thomas have?

Answer: ....

There were 100 students in a room. 70 students left the room and some new students entered. There were 120 students in the room in the end. How many new students entered the room?

Answer: ....

Oliver had a square field. He planted 6 trees on each side of the field with a tree on each corner. How many trees did he plant altogether?

Answer: ....

PROBLEM 46

Liz had some cakes. She ate 8 of them. Her friend Tom gave her 5 more cakes. Liz now had 20 cakes in all. How many cakes did Liz have at first?

Answer: ....

PROBLEM 47

Kate has 20 marbles and Bryan has 14 marbles. How many marbles should Kate give to Bryan so that both of them have the same number of marbles?

Answer: ………………………

PROBLEM 48

An ant crawled up 8 meters in the morning and 10 meters in the afternoon. At night, it crawled 3 meters back. How far was the ant from its starting point?

Answer: ………………………….

PROBLEM 49

Mia had 10 postcards. Anna gave her 22 postcards and Paul gave her 15. Mia then used 5 postcards to write to her friend. How many postcards were left with her in the end?

Answer: …………………………

There were 2 bottles of orange syrup, A and B. 2 liters of syrup was poured from bottle A to bottle B. 5 liters of syrup was then poured from bottle B to bottle A. In the end, bottle A and B contained 8 liters of orange syrup each.

a) How many liters of orange syrup was there in bottle A in the beginning.

b) How many liters of orange syrup was there in bottle B in the beginning?

Answer: ....

PROBLEM 51

There were 90 green marbles and 20 red marbles in a basket. Alex sold 40 green marbles and 10 red marbles. He then gave some marbles to his friend, and he had 20 marbles left. How many marbles did Alex give to his friend?

Answer: …………………………

PROBLEM 52

Bill has 95 seashells lesser than Joe. Joe has 45 seashells lesser than Ken. If Ken has 100 seashells, how many seashells do the three children have altogether?

Answer: ....

PROBLEM 53

There are a total of 86 people in Room A, Room B and Room C. There are 34 people in Room A. The number of people in Room C is 19 less than the number of people in Room A. How many people are there in Room B?

Answer: ………………………..

PROBLEM 54

Carl, Ann, and Bob have a total of 100 stamps. Carl has 45 stamps. Bob has 19 stamps fewer than Carl. How many stamps does Ann have?

Answer: ....

There were 70 ducks and 90 chicken on a farm. The farmer sold 40 ducks and 70 chicken. What was the total number of chicken and ducks left at the farm?

Answer: …………………………

PROBLEM 56

Rick has some marbles. He gave 8 marbles to Tom and bought 12 more marbles. Now, he has a total of 35 marbles. How many marbles did Rick have at first?

Answer: ………………………

PROBLEM 57

Jason had some stamps. He used 10 stamps for postage and bought another 28 stamps. Now, he has a total of 58 stamps. How many stamps did Jason have at first?

Answer: ....

PROBLEM 58

Alfred has $50 less than Bob.
Bob has $30 less than Mark.
If Mark has $100, find the total
amount of money Alfred and
Bob have.

Answer: ....

Ron weighs 90 kg. He is 22 kg heavier than Carol. Carol is 19 kg heavier than Sam. What is the total weight of Carol and Sam?

Answer: ....

Vivian saved $5 for 10 days. She wanted to buy a T- shirt that cost $10 and a pair of shorts that cost $25. How much money would Vivian have left after paying for the T- shirt and the pair of shorts from her savings?

Answer: …………………………

PROBLEM 61

Caroline sold 4 boxes of cupcakes. There were 5 cupcakes in each box. Felicia sold 9 cupcakes more than her. Each of them then had 14 cupcakes left. What was the total number of cupcakes both of them had at first?

Answer: …………………………

PROBLEM 62

There were 16 adults at a party. 8 of them ate 2 apple pies each and the rest ate 4 apple pies each. What was the total number of apple pies eaten?

Answer: …………………………

Samuel tied 5 bunches of 4 bananas each. Jake tied 7 bunches of 3 bananas each. How many more bananas did Jake tie than Samuel?

Answer: ....

PROBLEM 64

Maya bought 4 packets of biscuits. There were 6 biscuits in each packet. She ate 8 of the biscuits. How many biscuits were left?

Answer: ....

PROBLEM 65

Edward bought 5 packs of pens.
There were 12 pens in each pack.
He sold 24 of his pens. How
many pens were left?

Answer: ....

Sharon has 4 stickers. Thomas has 6 times as many stickers as Sharon. How many stickers do they have altogether?

Answer: ....

PROBLEM 67

Richard bought 5 trays of eggs. There were 12 eggs in each tray. 27 eggs fell and broke. How many eggs were not broken?

Answer: …………………………

PROBLEM 68

Henry has 4 sets of stamps. In each set, there were 12 stamps. He used 29 of these stamps. How many stamps did Henry have left?

Answer: ....

PROBLEM 69

Doris has five $2 notes. She has 8 times as many $5 notes as $2 notes. How many more $5 notes than $2 notes does she have?

Answer: …………………………

PROBLEM 70

Rebecca has 4 ribbons. Suzy has 7 times as many ribbons as Rebecca.

a) How many ribbons do they have altogether?

b) How many more ribbons does Suzy have than Rebecca?

Answer: ....

In a basket there were 7 apples. There are 5 times as many oranges as apples. How many more oranges than apples are there?

Answer: …………………………

PROBLEM 72

A puppy weighs 6 kilograms.
A dog weighs 6 times as much as
the puppy. How much heavier is
the dog than the puppy?

Answer: ....

PROBLEM 73

Mrs. Smith sold 8 jars of candies. Each jar of candies cost $4. How much money did Mrs. Smith earn from the candies she sold?

Answer: ....

PROBLEM 74

Tom has 8 marbles. Jack has 4 times as many marbles as Tom. How many marbles do they have altogether?

Answer: ....

PROBLEM 75

My favorite TV program has 3 advertisement breaks. Each advertisement lasts 3 minutes. How many minutes of advertisements would I sit through when I watch my favorite program?

Answer: ....

Mike's kitten drinks 2 liters of milk every day. If Mike buys 20 liters of milk, how much milk would he have left after a week?

Answer: ....

One watermelon is as heavy as
3 pineapples. One pineapple
is as heavy is as 3 pears. How
many pears are as heavy as 1
watermelon?

Answer: ………………………..

5 mugs of water can fill a pail and 4 cups of water can fill the same mug. How many cups of water are required to fill the same pail?

Answer: ....

PROBLEM 79

A basket has 4 red marbles and 5 yellow marbles. How many marbles are there in 6 such baskets?

Answer: ....

PROBLEM 80

There are 4 boys and 5 girls in a book shop. Each boy buys 3 books and each girl buys 4 books. How many books do they buy altogether?

Answer: ....

PROBLEM 81

Tom has 12 marbles. Jack has 4 times as many marbles as Tom. How many more marbles does Jack have than Tom?

Answer: …………………………

PROBLEM 82

Some guests were invited to a wedding dinner and were seated on 8 tables. There were 10 guests per table. 23 of the guests were women. 39 of the guests were men. How many children were at the dinner?

Answer: ....

Mrs. Lim bought 4 trays of eggs.
There were 10 eggs in each tray.
She used 12 of the eggs to bake
a birthday cake for her husband.
How many eggs were left?

Answer: ....

Mr. Smith bought 8 bottles of milk. Each bottle contained 2 liters of milk. His family consumed 5 liters of milk on Monday and 7 liters of milk on Tuesday. How much milk was left?

Answer: ....

In the market eggs are sold by dozens. Mrs. Roger bought 4 dozen eggs. She fried 9 of them for dinner and used 7 of them to bake a cake. How many eggs were left?

Answer: ....

PROBLEM 86

Richard bought 4 packets of oranges. There were 9 oranges in each packet. If he used 18 oranges, how many oranges did he have left?

Answer: …………………………

PROBLEM 87

Sandra sewed 5 buttons on a dress. After sewing 8 dresses, she had 12 buttons left. How many buttons did she have at first?

Answer: ....

Richard bought some bananas for his students. He gave 5 bananas to each of his 8 students and had 7 bananas left. How many bananas did Richard buy?

Answer: ....

PROBLEM 89

Alvin arranged 4 white tables and 5 red tables in a row. If he arranged 9 such rows, How many tables were there altogether?

Answer: …………………………

PROBLEM 90

Jason has 6 boxes of stamps. There were 12 stamps in each box. Bernice had 32 more stamps than Jason. How many stamps do they have altogether?

Answer: ....

PROBLEM **91**

There were 9 oranges in Bag A. There were twice as many oranges in Bag B than in Bag A. How many oranges were there altogether?

Answer: ....

PROBLEM 92

During an excursion, a school hired 4 big cars and 8 small cars. 8 people could sit in a big car and 6 in a small car How many people went for the excursion if all the cars went full?

Answer: ……………………………

PROBLEM 93

A packet of apples cost $5 and a packet of mangoes cost $4. Alicia had just enough money to buy 3 packets of apples and 7 packets of mangoes .How much money did she have?

Answer: ………………………

PROBLEM 94

There were 9 cars and 10 motorcycles in a car park. How many more car wheels were there than the motorcycle wheels?

Answer: ....

Jason baked some egg tarts and packed them into boxes of 5 each. He sold 8 such boxes and had 6 boxes left.

a) How many egg tarts did he bake?

b) How many egg tarts were left?

Answer: ....

PROBLEM 96

The number of students in Grade 2 can be divided into 5 groups of 20 students each. 15 students are Chinese, some are Americans and the remaining 19 students are Australians. How many American students are there Grade 2?

Answer: ....

PROBLEM 97

A fruit seller packed the fruits into 5 bags of 8 fruits each. 10 of these fruits were apples. There were 3 more pears than apples. The rest of the fruits were oranges. How many oranges were there at the fruit stall?

Answer: ....

PROBLEM 98

Kevin earned 10 stickers for reading books. He can exchange the stickers for the items written below. What can Kevin get with his 10 stickers?

1 sticker – bookmark

2 stickers – eraser

3 stickers – pencil

4 stickers – notepad

Answer: ………………………

Ariana has 5 baskets that hold a total of 15 pears. Each basket has a different number of pears. How many pears are there in each basket?

Answer: ....

PROBLEM 100

Jane bought 35 oranges and put them equally into 5 boxes. She gave away 3 boxes to her friends. How many oranges did she have left?

Answer: ....

PROBLEM 101

9 pair of shoes cost $54. What is the cost of 2 pair of shoes?

Answer: …………………………

PROBLEM 102

A mango tree has 8 mangoes. An orange tree has 10 oranges. An apple tree has 12 apples. Two children went picking the fruits and gets the same number of each kind of fruit. How many fruits in all will each child get if they picked all the fruits from all the trees?

Answer: ....

PROBLEM 103

There are some cows and ducks on a field that have a total of 76 legs. If there are 10 cows;

a) How many ducks are there?

b) How many cows and ducks are there altogether?

Answer: ....

I have 16 beautiful plants in my garden. My garden is divided into 4 equal parts. I have planted equal number of plants in each part. How many plants does each part have? How many plants are there in each part?

Answer: ....

PROBLEM 105

Mum calls her 5 children to sit at the table to have their snack. She has 12 crackers and she wants to give each child the same number of crackers. How many crackers does each child get? How many crackers are left?

Answer: ....

PROBLEM 106

A mother bought 12 cakes for her children. She wants to distribute them equally among her 2 children. How many does each child get?

Answer: …………………………

PROBLEM 107

Kevin reads 2 pages per hour.
How many hours will it take to
complete a 30 page book?

Answer: ....

PROBLEM 108

Lucy weighs 32 kg. Jerry weighs 2 kg lighter than Lucy but she is half as heavy as Sandra. How much does Sandra weigh?

Answer: ………………………

PROBLEM 109

A farmer had 260 chicken. He sold 130 in the morning and 90 of them in the afternoon. The rest of the chicken were put equally into 4 packets. How many chicken were there in each packet?

Answer: ....

PROBLEM 110

Jason bought 6 packets of mangoes. There were 12 mangoes in each packet. He sold the mangoes and put them into bundle of 3. How many bundles of mangoes were there?

Answer: ....

PROBLEM 111

Claire baked 5 trays of cupcakes. There were 8 cupcakes on each tray. She then gave 14 cupcakes to her children and packed the rest equally into 2 bags. How many cupcakes were there in each bag?

Answer: ....

PROBLEM 112

Amanda and her sister gathered all 98 of their toys and placed them on the shelves in their bedroom. If every shelf can carry a maximum of 7 toys, how many shelves will be filled?

Answer: ....

PROBLEM **113**

There was 20 kg of sugar.
Thomas used 14 kg and packed
the rest equally into small bags
of 2 kg each. How many such
bags of sugar did he have?

Answer: …………………………

PROBLEM 114

Kelvin had 8 packets of pears. There were 3 pears in each packet. He shared the pears with 6 friends. How many pears did each friend get?

Answer: ....

Max and Gavin had 28 oranges. Max ate 7 oranges and gave away 5 oranges. They then had the same number of oranges. How many oranges did Gavin have?

Answer: ....

PROBLEM 116

Tim has some marbles. There are 10 red marbles, 12 green marbles and 2 more orange marble than green marbles. If the marbles were put equally into 4 groups, how many marbles were there in each group?

Answer: …………………………

PROBLEM 117

A cake cost $4. Lara bought 4
cakes and gave the cashier
$60. How much change did she
get back?

Answer: …………………………

PROBLEM 118

At a stationery shop, whiteboard markers are sold at 3 for $2. How many markers can Daniel buy if he has $12?

Answer: …………………………

A pair of shoe cost $10 and a pair of glasses cost $26. During a sale, the price of all items fell by $2. How much would it cost to buy the shoe and the pair of glasses during the sale?

Answer: ………………………

PROBLEM 120

A pen and a pencil cost $6. Two similar pens cost $8. How much does the pencil cost?

Answer: …………………………

PROBLEM 121

A book cost $5, a pencil costs $2 and a pencil box costs $8. Jason wants to buy all the items but he needs $3 more. How much money does Jason have now?

Answer: …………………………

PROBLEM 122

Maggie bought 5 books for $10.
How much would it cost to buy 7
books?

Answer: ………………………….

PROBLEM 123

Anna bought a pen that cost $1.50. She gave the cashier $2. She was given ten-cent coins and twenty cent coins in change. What was the maximum number of twenty cent coins that Anna could have received?

Answer: ....

PROBLEM 124

A pair of socks cost $14 and a pair of shoes cost $18. Doris bought both items and gave the cashier a $60 note. She received a ten-dollar note and some two dollar notes in change. How many two dollar notes did she receive?

Answer: ....

PROBLEM 125

A watermelon cost as much as 3 bananas. Belinda spent $45 on 3 watermelons. What was the cost of one banana?

Answer: ....

PROBLEM 126

A pen cost $6. It cost $3 more than a sharpener. Maya had $42. She bought 2 pens and 2 sharpeners. How much money was left with Maya?

Answer: ....

PROBLEM 127

Tim had 6 two-dollar notes, 4 five-dollar notes and 2 one-dollar notes. He spent $20 on a book and $6 on a magazine. How much money was left with Tim?

Answer: …………………………

PROBLEM 128

3 oranges cost as much as 3 apples. If an orange costs $2, what is the cost of an apple?

Answer: ………………………

PROBLEM 129

A kiwi fruit and 3 mangoes cost $7. If a mango cost $2, what is the cost of one kiwi fruit?

Answer: ....

Samuel spent $3 on a pair of socks. He spent $14 more on a scarf than on the pair of socks. If he had $2 left, how much did he have at first?

Answer: ....

PROBLEM 131

Henry earns $10 for every half an hour that he works. If he works from 8.30 am to 12. 30 am, how much would he earn altogether?

Answer: ………………………

PROBLEM 132

Max gets paid $6 an hour for cleaning his neighbor's garden. On Saturday, he cleaned from 2:00 pm to 5:00 pm. How much money did he make?

Answer: ....

PROBLEM 133

A book costs $6. Bryan bought 6 such books and gave the cashier a $50 note. The cashier gave him the change in two-dollar notes. How many two-dollar notes did he get in change?

Answer: ………………………….

Kevin bought 8 notebooks. Each notebook cost $2. He paid for the notebooks with a $50 note. How much change did he receive?

Answer: ....

PROBLEM 135

A magazine costs $4. Jennifer bought 7 such magazines and had $8 left. How much money did she have at first?

Answer: ....

PROBLEM 136

Paul was queuing for food in a restaurant. There were 4 children in front of him and 3 children behind him. Each of them had $4 each. How much money did the children have altogether?

Answer: ....

PROBLEM 137

Dennis bought a watch for $86 and a pen for $12. He gave the cashier a $100 note and received his change in twenty cent coins. How many twenty cent coins did he receive?

Answer: ....

PROBLEM 138

Jessica had $45. She spent $18 on a toy and the rest of the money on 3 similar T-shirts. How much did each T-shirt cost?

Answer: ....

PROBLEM 139

Kevin needs to earn 9 dollars. He earns 1 dollar and 50 cents an hour for working at the library. He has already worked for 4 hours. How much longer should Kevin work to earn the money he needs?

Answer: ....

James gets 80 cents for his pocket money allowance each Saturday. On Monday, he bought a pencil for 10 cents. On Tuesday, he spent 15 cents on candy. On Wednesday, he wanted to buy a book that cost 30 cents. Did he have enough money to buy the book?

Answer: ....

PROBLEM 141

An adult ticket to a concert costs $8 and a child's ticket is $5. How much do 2 adults and a child have to pay to go to the concert?

Answer: …………………………

Kelvin paid $52 for a pair of shoe. This was $14 less than what he paid for a jacket. How much did he pay altogether?

Answer: …………………………

PROBLEM 143

After buying some stickers for $30.00, Sandra has $7.00 left. How much money did Sandra have to begin with?

Answer: ....

PROBLEM 144

Carol has $45.00 and Lucy has $8.00. How much more does Carol have than Lucy?

Answer: ....

PROBLEM 145

Mark receives $15 every month for his pocket money allowance. He puts $7 out of it into a piggy bank until his piggy bank has $119. How many months has he been saving part of his pocket money?

Answer: ....

Mr. Carter collected $6 each from his students for their upcoming field trip. All of his students went on the field trip and he collected $192 in all. How many students are there in Mr. Carter's class?

Answer: ....

PROBLEM 147

Carmen gets $20 every week for her lunch allowance. She sets aside $2 every day. How many days would it take her to save up $60?

Answer: ....

PROBLEM 148

Emily is selling concert tickets for $4. She collects a total of $284. How many tickets did she sell?

Answer: ....

PROBLEM 149

Laura spent $150 at a jewellery store and $50 at a department store. She had $90 left. How much did she have at first?

Answer: ....

PROBLEM 150

Melissa bought a shoe that cost $18 and received 2 two-dollar notes in change. How much did she have at first?

Answer: ....

PROBLEM 151

Julia's monthly pocket money allowance is $5.00. She baby sat her neighbor's kid and earned an extra $1.75. She went to the store with her friend and bought an ice-cream for $2.25. On her way home she found $1.00. How much money did Julia have?

Answer: …………………………

PROBLEM 152

Jacob got $25.00 for his birthday as a gift from his grandmother. He spent $9.00 from it on buying a book. Jacob then decided to buy a remote control car for $12.00 and a fizzy soda for $2.50. How much money does Jacob have left?

Answer: ....

PROBLEM 153

Louis wants to buy a comic book and it costs $13.00. He has $6.85. How much more money does Louis need to buy the comic book?

Answer: ....

PROBLEM 154

Amy gave the cashier a ten-dollar note. The cashier returned her 4 two-dollar notes, 3 twenty-cent coins and 1 ten-cent coin for change. How much money did she spend?

Answer: ....

160

Copyright © 2017 Learn 2 Think

PROBLEM 155

Keith bought a book for $8.35 and a bag of pens for $3.25. He used a $20 bill to pay for them. How much change did Ken get?

Answer: ....

PROBLEM 156

Mary used all her money to buy 4 boxes of colored pens. A box of colored pens cost $3 less than a box of crayons. If one box of crayons cost $8, how much money did Mary have at first?

Answer: ………………………

PROBLEM 157

Thomas and Mary bought a gift together for their best friend Lucy. Thomas paid $3 more than Mary. If the gift cost $21, how much did Thomas pay for the gift?

Answer: …………………………

PROBLEM 158

There were 6 apple pies and 8 chocolate pies in a box. An apple pie cost $5 and a chocolate pie cost $4. How much money was earned from sale of all the pies?

Answer: …………………………

Mark bought a table and 4 chairs for $200. He paid $176 for the table.

a) Find the cost of 4 chairs

b) Find the cost of each chair.

Answer: ……………………….

Mike bought 8 oranges. Each orange cost $3. He paid for the oranges with a $100-note. How much change did he get back?

Answer: ....

PROBLEM 161

A pencil cost $3. A pen cost $4 more than the pencil. Alan bought 5 pens. He gave the cashier four $10-notes. How much change did he get back?

Answer: ……………………………

A papaya cost $2. A watermelon cost $3 more than the papaya. Mr James bought 8 watermelons. He gave the fruit seller a $100 note. How much change did he receive?

Answer: ……………………………

PROBLEM 163

A diary cost $3. A dictionary cost $5 more than the diary. George bought 4 dictionaries. He gave the cashier a $50-note. How much change did he receive?

Answer: ....

PROBLEM 164

A blouse costs $8. A jacket costs thrice as much the blouse. What is the total cost of the blouse and the jacket?

Answer: ....

PROBLEM 165

Jane went shopping with three $50 notes. She bought a pen which cost $20 and a dress which cost $40. How much money did she have left?

Answer: …………………………

Kevin went shopping with eighteen $2 notes. He spent $29 on a jacket. How much money was left with Kevin?

Answer: ………………………

PROBLEM 167

A packet of 3 apples cost $1.50 and a packet of 4 mangoes cost $2. How much would it cost to buy 3 apples and 6 mangoes?

Answer: …………………………

Daniel wishes to buy 7 mangoes. Each mango costs $5. He only has $21 with him. How much more money does he need?

Answer: ....

A toy car costs $32. A stuffed toy costs $28 less than the stuffed toy. Tony bought 9 stuffed toys. He gave the cashier two $20 notes. How much change did he get back?

Answer: ....

PROBLEM 170

A DVD cost $20. A book cost $15 less than the DVD. Jack bought 9 books and gave the cashier a $50 note. How much change did he get back?

Answer: ....

PROBLEM 171

A watch cost $80. A belt cost $70 less than the watch. Mary bought 5 belts. She gave the cashier a $100-note. How much change did she receive?

Answer: ....

PROBLEM 172

Tom had some money that he wanted to use for shopping. He spent $25 on a jacket and withdrew another $80 from the ATM. Now, he has a total of $200. How much money did Tom have at first?

Answer: ....

PROBLEM 173

If I played soccer at 4:00 o'clock and ate dinner 3 hours later, what time did I eat my dinner?

Answer: …………………………

Kevin started eating his lunch at 1:05pm and finished at 2:05pm. For how long did Kevin eat his lunch?

Answer: ....

PROBLEM 175

If your school begins at 9:30 am and gets over at 3:30 pm, how many hours are you at school every day?

Answer: ....

PROBLEM 176

The first day of a certain month that has 30 days is a Monday. How many Mondays does this month have?

Answer: ....

PROBLEM 177

A concert started at 8.30 pm and ended at 10.00 pm. How long did the concert run?

Answer: ....

PROBLEM 178

Samuel is 8 years old this year. His brother is 3 years older than him. How old will his brother be in 5 years time?

Answer: ....

PROBLEM 179

Tracy is 36 years old. Three years ago, her daughter was 9 years old. How old will Tracy be when her daughter is 20 years old?

Answer: ....

PROBLEM 180

Tom is 12 years old, His mother is 38 years old. How old will Tom be when his mother is 60 years old?

Answer: ....

PROBLEM 181

Ted is 18 years old. His father is 44 years old. How old will Ted be when his father is 90 years old?

Answer: …………………………

PROBLEM 182

Susan is 47 years old. Her husband is 8 years older than her. What will be their total age in 12 years time?

Answer: ………………………….

PROBLEM 183

John is 82 years old. His grandson is 59 years younger than him. What will be their total age in 14 years time?

Answer: ....

PROBLEM 184

Alfred is 57 years old. His wife is 9 years younger to him. What was their total age 12 years ago?

Answer: ....

190

PROBLEM 185

Tom is 16 years younger than his mother. His father is 7 years older than his mother. How much older is his father than Tom?

Answer: …………………………

Josephine is 24 years older than Ron. Ron is 67 years younger than Jake. How much older is Jake than Josephine?

Answer: ....

PROBLEM 187

Kelvin is 4 years old this year. His brother is thrice as old as him. How old was his brother three years ago?

Answer: ....

Amelia bought some flour. She used 3 kg to bake cupcakes and packed the rest equally into 4 bags of 2 kg each. How much flour did she have at first?

Answer: …………………………

PROBLEM 189

4 bags of potatoes and 3 bags of rice have a mass of 45 kg. If each bag of rice had a mass of 7 kg, what is the mass of each bag of potatoes?

Answer: ....

PROBLEM 190

An elephant weighs 28 kg more than a deer. A deer weighs 45 kg less than the tiger. Which animal is heavier; the tiger or the elephant?

Answer: ....

PROBLEM 191

Terry is 18 kg lighter than Jason.
Rick is 19 kg heavier than Jason.
Who is heavier among Terry and
Jason?

Answer: …………………………

There was 16 kg of flour in bag A. After some flour was poured from Bag A to Bag B, there was 4 kg more flour in bag A than in bag B. There was 9 kg of flour in bag B. How much flour was poured from bag A to bag B?

Answer: …………………………

PROBLEM 193

A mango is 30 grams heavier than a kiwi fruit. If the mango weighs 60 grams, how much do 3 similar kiwi fruits weigh?

Answer: ....

PROBLEM 194

Timothy is 10 kg heavier than Sam. Sam is 4 kg heavier than Andy. If Andy is 24 kg, what is the total mass of the 3 boys?

Answer: ....

PROBLEM 195

A rope is 100 meters long. It is cut into 2 pieces. If one piece is 32 meters long, what is the length of the other piece?

Answer: ....

PROBLEM 196

A box filled with toys weighs 22 kg. The toys weigh 19 kg. How much do 5 such empty boxes weigh?

Answer: ....

PROBLEM 197

Joe weighs 78 kg. Rick weighs 19 kg lighter than Joe. Leon is 24 kg heavier than Rick. Find the total weight of the three men.

Answer: ....

PROBLEM 198

A laptop weighs 2 kg. A table is 6 times as heavy as a laptop.

a) What is the total weight of the two items?

b) How much heavier is the table than the laptop?

Answer: ....

PROBLEM 199

A mug contains 50 milliliters of water. It contains 45 milliliters less water than a jug. The jug contains 60 milliliters more water than a cup. What is the total amount of water in the 3 containers?

Answer: ....

PROBLEM 200

Mary has two cows in her house. Cow A gives 2 liter milk per day and Cow B gives 5 liter milk per day. How many more liters milk does cow B give than cow A?

Answer: ....

SOLUTIONS

 ## Solution to Question 1

The sum of the two digits is 12.
Possible combinations of numbers for the sum to be 12 are:
1, 11
2, 10
3, 9
4, 8
5, 7
6, 6
The product of the two digits is 27
1 x 11 = 11
2 x 10 = 20
3 x 9 = 27
Therefore the no. is 39.

 ## Solution to Question 2

Alison is reading a book.
She noted that the sum of the 2 consecutive pages that she is reading now is 35.
Lets try some combinations of consecutive numbers that can give us 35 when added.(Guess and check method)

15 + 16 = 31
16 + 17 = 33
17 + 18 = 35
The page numbers of these two pages are 17 & 18.

 ## Solution to Question 3

On a farm there are a total of 7 goats and hens
Lets try some combinations that give 24.
3 goats + 4 hens = 3 x 4 + 4 x 2 = 12 + 8 = 21
4 goats + 3 hens = 4 x 4 + 3 x 2 = 16 + 6 = 22

208

5 goats + 2 hens = 5 x 4 + 2 x 2 = 20 + 4 = 24 Number of goats and hens on the farm are 5 & 2.

Solution to Question 4

The number is more than 25 but less than 40.
Ones digit is twice the tens digit
Possible numbers are
1, 2
2, 4
3, 6
4, 8
The digit adds up to an odd number and number is between 25 and 40.
The digits add up to 3 + 6 = 9 (odd number)
Therefore the number is 36.

Solution to Question 5

The sum of 3 odd numbers is 21.
Let's write down the odd numbers less than 21.
1, 3, 5, 7, 9, 11, 13, 1, 5, 17, 19
Possible numbers are :
1 + 5 + 15 = 21
1 + 3 + 17 = 21
1 + 7 + 13 = 21
1 + 9 + 11 = 21
3 + 5 + 13 = 21
3 + 7 + 11 = 21

Solution to Question 6

The difference between two number is 142.

The bigger number is 196.
The smaller number is = 196 − 142 = 54

 Solution to Question **7**

I am greater than 6 and less than 16.
Numbers between 6 & 16 are
7, 8, 9, 10, 11, 12, 13, 14, 15
I am an odd number.
Therefore odd numbers between 6 & 16 are
7, 9, 11, 13, 15
When you count by 5's, you say my name.
Therefore number which can be counted by 5's is 15.

 Solution to Question **8**

The largest 4-digit odd number which has no repeating digits is 9875

 Solution to Question **9**

Four children are lining up to buy lunch.
Their names are Lucy, Bill, Jim, and Mary.
Lucy is first.
Bill is last.
Mary is behind Lucy and in front of Jim.
Order in which the four children are standing:
Lucy   Mary   Jim   Bill
Child third in line is Jim.

 Solution to Question **10**

10 people were in a queue to return their library books.

Joanne was the 7th person from the front.
Therefore Joanne position from the back = 10 − 7 + 1 = 4th
Bob was the 6th person from the back.
Number of people standing between Bob and Joanne = 1

Solution to Question 11

20 students are lining up to board the bus.
Jack is in 12th place from the front. This means his position is 9th from the back.
Ann is at the 5th place from the back. Number of people standing between Jack and Ann = 3

Solution to Question 12

20 books are placed on a stack.
The 8th book from the top is a dictionary.
Position of dictionary from bottom = 20 − 8 + 1 = 13th
The 16th book from the bottom is a phone directory.
Number of books between the dictionary and the phone directory = 2

Solution to Question 13

Charles is in the 5th place in a queue to buy movie tickets.
Tony is last in the queue.
There are 5 people in between Tony and Victor and 2 people between Charles and Victor.
Position of Victor in the queue from front = 5 + 2 + 1 = 8
Number of people in the queue = 8 + 5 + 1 = 14

 Solution to Question **14**

Doreen is in the 7th place in a queue to buy entry tickets to a zoo.
Xavier is the last in the queue.
There are 8 people between Xavier and Ronald and 5 people in between Doreen and Ronald.
a) Position of Ronald in the queue from front = 7 + 5 + 1
(including Ronald) = 13
b) Number of people in the queue = 13 + 8 + 1 = 22

 Solution to Question **15**

20 athletes are running one behind the other along a jogging track. The runner in the 15th position is Tom.
Tom's position from back = 20 − 15 + 1 (adding 1 to count Tom = 6th
The 12th runner from the back is Fred.
Number of runners between Tom and Fred = 5 (remember to exclude Tom and Fred while counting the runners between them.)

 Solution to Question **16**

10 trees are planted along a straight road.
The 8th tree from the front is a palm tree.
Palm tree from back = 3rd
The 5th tree from the back is a mango tree.
Number of trees planted in between the palm tree and the mango tree
= 1

 Solution to Question **17**

There are 120 red marbles and 130 blue marbles in a box.
Another 200 green marbles are added into the box.

212

Total Number marbles in the box now = 120 + 130 + 200 = 450

 Solution to Question **18**

Paul read 20 pages of a magazine on Sunday.
He read 14 more pages on Monday than on Sunday = 20 + 14 = 34
There were 40 unread pages.
Number of pages in the magazine = 20 + 34 + 40 = 94

 Solution to Question **19**

There were 196 pages in a book.
Jason reads 18 pages on Monday
28 pages on Tuesday
38 pages on Wednesday
Number of pages read each day is increasing by 10.
Total pages read = 18 + 28 + 38 = 84
On Thursday pages read = 84 + 48 = 132
On Friday pages read = 132 + 58 = 196
Total Number of days he would take to finish reading the entire
book in this pattern = 5

 Solution to Question **20**

Janice has 120 mangoes and 240 tomatoes.
Total = 120 + 240 = 360
Janice has 82 mangoes fewer than Karen.
Number of mangoes Karen has = 120 + 82 = 202
The total number of mangoes and tomatoes that Karen and Janice had
= 360 + 202
= 562 Answer

Solution to Question 21

Bryan sold 120 chicken on Monday and 200 chicken on Tuesday.
Total Number of chicken sold = 120 + 200 = 320
He packed the rest of his chicken equally into 10 boxes of 10 chicken each.
Total chicken in 10 boxes = 10 x 10 = 100
Total Number of chicken he had at first = 320 + 100 = 420

 ## Solution to Question 22

Florence used 16 bananas to bake some cakes and 14 bananas to bake some cookies.
Bananas used = 16 + 14 = 30
She had 62 bananas left.
Total bananas = 30 + 62 = 92
Number of bananas she had at first = 92

 ## Solution to Question 23

Kathy has 40 stamps.
Sharon has 18 more stamps than Kathy = 40 + 18 = 58
Joe has 37 more stamps than Sharon = 58 + 37 = 95
Number of stamps that Joe and Sharon had altogether
= 95 + 58 = 153

 ## Solution to Question 24

A clock chimes once at 1 o'clock, twice at 2 o' clock, thrice at 3 o'clock.
Number of times it would chime by 7 o' clock
= 1 + 2 + 3 + 4 + 5 + 6 + 7 = 28

214

 **Solution to Question 25**

The largest number that can be written using the digits 4, 8, 5 is
854

 **Solution to Question 26**

There were 130 cars in a car park.
20 cars drove out of the car park and 42 cars drove in.
Number of cars in the car park in the end = 130 − 20 + 42 = 152

 **Solution to Question 27**

There were 120 people in a shopping center.
38 of them were children.
20 of them were women and the rest were men.
Men = 120 − 20 − 35 = 65
Number of men more than women in the shopping center
= 65 − 20 = 45

 **Solution to Question 28**

Kevin had some stickers in the beginning.
His mother gave him 14 stickers. After he gave 18 stickers to Jerry
he had 70 stickers left.
Number of stickers Kevin had in the beginning + 14 − 18 = 70
Number of stickers Kevin had in the beginning − 4 = 70
Number of stickers Kevin had in the beginning = 70 + 4 = 74
Number of stickers Kevin had at first = 74

Solution to Question 29

Sam and Alex had 20 marbles altogether.
Sam gave Alex 4 marbles and now they have an equal number of marbles.
Lets try various combinations of marbles Sam and Alex had to have 20 marbles altogether.

    a) 15 + 5 = 20
    b) 12 + 8 = 20
    c) 14 + 6 = 20

Only the condition c) is possible = 14 − 4 = 6 + 4 = 10
Number of marbles Sam had in the beginning = 14

 **Solution to Question 30**

There were 120 passengers on a train.
At the first stop, 14 passengers got on and 50 passengers got off.
= 122 − 14 + 50
=158
At the second stop, some passengers got off the train but none got on.
There were then 24 passengers left on the train.
The number of passengers who got off the train at the second stop
= 158 − 24
= 134

Solution to Question 31

There were 160 potatoes in a box.
18 rotten ones were thrown away.
Remaining = 160 − 18 = 142
Some potatoes were then baked.
120 potatoes were left in the box.
Number of potatoes that were baked = 142 − 120 = 24

216

 Solution to Question **32**

A box contains 40 cupcakes.
There are 16 strawberry cupcakes, 22 blueberry cupcakes and some chocolate cupcakes.
Number of chocolate cup cakes = 40 − 16 − 22 = 2
Number of chocolate cupcakes in 6 such boxes = 6 x 2 = 12

 Solution to Question **33**

Henry had 70 stickers.
Jane had 95 stickers more than him = 70 + 95 = 165
Patrick had 15 stickers fewer than Janice = 165 − 15 = 150
Number of stickers they had altogether = 70 + 165 + 150 = 385

 Solution to Question **34**

Sam had some coins.
He gave away 14 coins to Gerald.
His father later gave him 18 more coins.
He finally has a total of 30 coins.
Number of coins Sam had at first - 14 + 18 = 30
Number of coins Sam had at first + 4 = 30
Number of coins Sam had at first = 30 − 4
Number of coins Sam had at first = 26

 Solution to Question **35**

Karen has 48 Australia stamps.
She has 14 fewer England stamps than Australia stamps.
England stamps = 48 − 14 = 34
Number of stamps she had altogether = 48 + 34 = 82

217

 Solution to Question **36**

Jennifer had 52 stamps.
She gave 34 stamps to her brother and bought some new stamps. She had 74 stamps in the end.
52 − 34 + Number of stamps bought = 74
Number of stamps bought + 18 = 74
Number of stamps bought = 74 − 18
Number of stamps bought = 56

 Solution to Question **37**

David had 130 pears.
He threw away 85 rotten ones and bought another 72 pears.
Number of pears he had in the end = 130 − 85 + 72 = 117

 Solution to Question **38**

Mr. Ted bought 7 packs of pencils for his class. The total cost of pencils is $98.
Every pack of pencils cost the same amount.
Cost of all the packs = 7 + 7 + 7 + 7 + 7+ 7 + 7 + 7 + 7+ 7 + 7 + 7 + 7 + 7 = $98
Cost of each pack = $7

 Solution to Question **39**

Susan had 130 stamps.
She gave some stamps to her brother and 26 stamps to her sister.
She had 120 stamps left.
Stamps with the brother + 26 + 120 = 430
Stamps with the brother + 146 = 430
Stamps with the brother = 430 − 146 = 284

218

No of stamps she gave to her brother = 284

 **Solution to Question 40**

There were 450 people at a party.
There were 130 men and 120 women.
Total Number of adults = 130 + 120 = 250
The rest were children.
Number of children in the party = 450 − 250 = 200

 Solution to Question 41

Tim had 36 toy cars.
Jason had 20 more toy cars than him = 20 + 36 = 56
Kevin had 50 fewer toy cars than Jason = 56 − 50 = 6
Number of toy cars they had altogether = 36 + 56 + 6 = 98

 Solution to Question 42

Mark had 80 bananas.
Simon has 20 bananas lesser than Mark = 80 − 20 = 60
Sarah had 15 bananas more than Simon = 60 + 15 = 75
Sarah sold 30 bananas.
Number of bananas left with Sarah= 75 − 30 = 45

 Solution to Question 43

Jessica, Mary and Thomas had 150 chocolates altogether.
Jessica had 60 chocolates.
Mary had 10 chocolates more than Jessica = 60 + 10 = 70
Total Number of chocolates with Mary and Jessica = 60 + 70 = 130
Number of stamps with Thomas = 150 − 130 = 20

219

 **Solution to Question 44**

There were 100 students in a room.
70 students left the room and some new students
entered. There were 120 students in the room in the end.
100 − 70 + Number of students entered = 120
30 + Number of students entered = 120
Number of students entered = 120 − 30 = 90
Number of students who entered the room = 90

 **Solution to Question 45**

Oliver had a square field.
He planted 6 trees on each side of the field with a tree on each
corner. Number of trees in corner = 4
Number of trees on one side = 6 − 2 = 4
Number of trees on 4 sides = 4 x 4 = 16
Number of trees he planted altogether = 16 + 4 = 20

 **Solution to Question 46**

Thomas gave Liz 5 cakes and she ate 8 cakes.
Finally she had 20 cakes.
Number of cakes with Liz at first + 5 − 8 = 20
Number of cakes with Liz at first − 3 = 20
Number of cakes with Liz at first = 20 + 3 = 23
Number of cakes Liz had at first = 23

Solution to Question 47

Kate has 20 marbles and Bryan has 14 marbles.
Total Number of marbles = 20 + 14 = 34
For Kate and Bryan to have same number of marbles each one of

them should have 34/2 = 17 marbles
Number of marbles Kate should give to Bryan so that both of them have the same number of marbles = 20 − 17 = 3

 ## Solution to Question 48

An ant crawled up 8 meters in the morning and 10 meters in the afternoon.
Total distance crawled = 8 + 10 = 18 meters
At night it crawled back 3 meters.
Distance by which the ant is away from its starting point
= 18 − 3 = 15 meters

 ## Solution to Question 49

Mia had 10 postcards.
Anna gave her 22 postcards and Paul gave her 15 postcards = 10 + 22 + 15 = 47
Mia then used 5 postcards to write to her friends.
Number of postcards left with her = 47 − 5 = 42

 ## Solution to Question 50

There were 2 bottles of orange syrup, A and B.
In the end, bottle A and B contained 8 liters of syrup each.
Total syrup = 8 x 2 = 16 liters
Let us work backwards. This means whatever quantity of syrup was poured out will be added back to the final amount of syrup each bottle had.
2 liters of syrup was poured from bottle A to bottle B.

Bottle A	Bottle B
8 + 2 = 10	8 − 2 = 6

5 liters of syrup was then poured from bottle B to bottle A.

$10 - 5 = 5 \qquad 6 + 5 = 11$

Number of liters of syrup in bottle A = 5 liters

Number of liters of syrup in bottle B = 11 liters

 Solution to Question **51**

There were 90 green marbles and 20 red marbles in a basket.

Alex sold 40 green marbles and 10 red marbles.

Total green marbles left after being sold = 90 − 40 = 50

Total red marbles left after being sold = 20 − 10 = 10

Total number of remaining marbles = 50 + 10 = 60

He then gave some marbles to his friend.

He had 20 marbles left.

Number of marbles Alex gave to his friend = 60 − 20 = 40

 Solution to Question **52**

Ken has 100 seashells.

Joe has 45 seashells less than Ken = 100 − 45 = 55

Bill has 15 seashells less than Joe = 55 − 15 = 40

Number of seashells, the three children had altogether

= 100 + 55 + 40 = 195

 Solution to Question **53**

There are a total of 86 people are in Room A, Room B and Room C.

There are 34 people in Room A.

The number of people in Room C is 19 less than the number of
people in Room A = 34 − 19 = 15

Number of people in Room A & Room C = 34 + 15 = 49

Number of people in Room B = 86 − 49 = 37

 ## Solution to Question 54

Carl, Ann, and Bob have a total of 100 stamps.
Carl has 45 stamps.
Bob has 19 fewer stamps than Carl = 45 − 19 = 26
Number of stamps Ann had = 100 − 45 − 26 = 29

 ## Solution to Question 55

There were 70 ducks and 90 chicken on a farm.
The farmer sold 40 ducks and 70 chicken.
Remaining chicken = 90 − 70 = 20
Remaining ducks = 70 − 40 = 30
Total number of chicken and ducks left on the farm = 20 + 30 = 50

 ## Solution to Question 56

Rick has some marbles.
He gave 8 marbles to Tom and bought 12 more marbles. Now he
has a total of 35 marbles.
Number of marbles Rick had at first− 8 + 12 = 35
Number of marbles Rick had at first + 4 = 35
Number of marbles Rick had at first = 35 − 4
Number of marbles Rick had at first = 31

 ## Solution to Question 57

Jason had some stamps. He used 10 stamps for postage and
bought another 28 stamps.
Now Jason has a total of 58 stamps.
Number of stamps Jason had in the beginning − 10 + 28 = 58
Number of stamps Jason had at first = 40

 **Solution to Question 58**

Mark has $100.
Bob has $30 less than Mark = 100 − 30 = $70
Alfred has $50 less than Bob = 70 − 50 = $ 20
Total amount of money Alfred and Bob have = 70 + 20 = $90

 **Solution to Question 59**

Ron weighs 90 kg.
He is 22 kg heavier than Carol.
Carol weighs = 90 − 22 = 68 kg
Carol is 19 kg heavier than Sam.
Sam weighs = 68 − 19 = 49 kg.
The total weight of Carol and Sam = 68 + 49 = 117 kg

 **Solution to Question 60**

Vivian saved $5 for 10 days.
Over 10 days the amount of money she saved = 10 x 5 = $50
She wanted to buy a T-shirt and a pair of shorts.
The T-shirt cost $10 and the pair of short cost $25.
Total cost of both = 10 + 25 = $35
Amount of money left with Vivian after paying for the T-shirt
and shorts = 50 − 35 = $15

 **Solution to Question 61**

Caroline sold 4 boxes of cupcakes.
There were 5 cupcakes in each box.
Number of cupcakes Caroline sold = 5 x 4 = 20 Felicia sold
9 cupcakes more than Caroline = 20 + 9 = 29
Each of them then had 14 cupcakes left.

224

Total number of cupcakes both of them had at first
= 20 + 29 + 14 + 14 = 77

 ## Solution to Question 62

There were 16 adults at a party.
8 of them ate 2 apple pies each.
The rest ate 4 apple pies each
The total number of apple pies eaten = 8 x 2 + 8 x 4 = 16 + 32 = 48

 ## Solution to Question 63

Samuel tied 5 bunches of 4 bananas each.
Total number of bananas Samuel tied = 5 x 4 = 20
Jake tied 7 bunches of 3 bananas each.
Total number of bananas tied = 7 x 3 = 21
Number of more bananas Jake tied than Samuel = 21 − 20 = 1

 ## Solution to Question 64

Maya bought 4 packets of biscuits.
There were 6 biscuits in each packet.
Total biscuits = 6 x 4 = 24
She ate 8 of the biscuits.
Number of biscuits left = 24 − 8 = 16

 ## Solution to Question 65

Edward bought 5 packs of pens.
There were 12 pens in each pack.
Total pens = 12 x 5 = 60
He sold 24 of his pens.

Number of pens left = 60 − 24 = 36

 ## Solution to Question 66

Sharon has 4 stickers.
Thomas has 6 times as many stickers as Sharon = 6 x 4 = 24
Number of stickers they had altogether = 24 + 4 = 28

 ## Solution to Question 67

Richard bought 5 trays of eggs.
There were 12 eggs in each tray.
Total eggs = 12 x 5 = 60
27 eggs fell and broke.
Number of eggs that are not broken = 60 − 27 = 33

 ## Solution to Question 68

Henry has 4 sets of stamps.
In each set, there were 12 stamps.
Total Number of stamps = 12 x 4 = 48
He used 29 of these stamps.
Number of stamps Henry had left = 48 − 29 = 19

 ## Solution to Question 69

Doris has five $2 notes.
She has 8 times as many $5 notes as $2 notes = 8 x 5 = 40 notes
Number of more $5 notes than $2 notes she had = 40 − 5 = 35

226

Solution to Question 70

Rebecca has 4 ribbons.
Suzy has 7 times as many ribbons as Rebecca = 7 x 4 = 28
Ribbons they have altogether = 28 + 4 = 32
Number of more ribbons Suzy has than Rebecca = 28 − 4 =
24

Solution to Question 71

In a basket there were 7 apples.
There are 5 times as many oranges as apples = 7 x 5 = 35
Number of more oranges than apples = 35 − 7 = 28

Solution to Question 72

A puppy weighs 6 kilograms.
A dog weighs 6 times as much as the puppy = 6 x 6 = 36 kilograms
Number of kilograms by which the dog is heavier than the puppy
= 36 − 6 = 30 kilograms

Solution to Question 73

Mrs. Smith sold 8 jars of candies.
Each jar of candies cost $4.
Money she received from the candies she sold = 8 x 4 = $32

Solution to Question 74

Tom has 8 marbles.
Jack has 4 times as many marbles as Tom = 8 x 4 = 32
Number of marbles they have altogether = 32 + 8 = 40

 **Solution to Question 75**

TV program has 3 advertisements breaks.
Each advertisement lasts 3 minutes.
Number of minutes of advertisements I would sit through
while I watch my favorite program = 3 x 3 = 9 minutes

 **Solution to Question 76**

Mike's kitten drinks 2 liters of milk every day.
Amount of milk kittens will drink over a week = 7 x 2 = 14
Mike buys 20 liters of milk.
Amount of milk remaining after a week = 20 − 14 = 6 liters

 **Solution to Question 77**

One watermelon is as heavy as 3 pineapples.
One pineapple is as heavy is as 3 pears.
3 pineapples = 3 x 3 = 9 pears
Number of pears as heavy as 1 watermelon = 9

 **Solution to Question 78**

5 mugs of water can fill a pail.
4 cups of water can fill the same mug.
Number of cups of water required to fill the same pail = 5 x 4 = 20

 **Solution to Question 79**

A basket has 4 red marbles and 5 yellow marbles.
Total marbles in a basket = 4 + 5 = 9
Number of marbles in 6 such basket = 9 x 6 = 54

228

 Solution to Question **80**

There are 4 boys and 5 girls in a book shop.
Each boy buys 3 books.
Number of books 4 boys buy = 4 x 3 = 12
Each girl buys 4 books.
Number of books 5 girls buy = 5 x 4 = 20
Number of books they bought altogether = 12 + 20 = 32

 Solution to Question **81**

Tom has 12 marbles.
Jack has 4 times as many marbles as Tom = 4 x 12 = 48
Number of more marbles Jack has than Tom = 48 − 12 = 36

 Solution to Question **82**

Some guests were invited to a wedding dinner and were seated on
8 tables.
There were 10 guests per table.
Total Number of guests = 10 x 8 = 80
23 of the guests were women.
39 of the guests were men.
Number of children at the dinner = 80 − 23 − 39 = 18

 Solution to Question **83**

Mrs. Lim bought 4 trays of eggs.
There were 10 eggs in each tray.
Total Number of eggs = 4 x 10 = 40
She used 12 of the eggs to bake a birthday cake for her husband.
Number of eggs left = 40 − 12 = 28

 **Solution to Question 84**

Mr. Smith bought 8 bottles of milk.
Each bottle contained 2 liters of milk.
Total milk = 8 x 2 = 16 liters
His family consumed 5 liters of milk on Monday and 7 liters of milk on Tuesday
Consumed milk = 5 + 7 = 12 liters
Milk left = 16 − 12 = 4 liters

 **Solution to Question 85**

In the market eggs are sold by dozens.
1 dozen = 12
Mrs. Roger bought 4 dozen eggs.
Total no. of eggs = 12 x 4 = 48
She fried 9 of them for dinner and used 7 of them to bake a cake.
Used eggs = 9 + 7 = 16
No. of eggs left = 48 − 16 = 32

 **Solution to Question 86**

Richard bought 4 packets of oranges.
There were 9 oranges in each packet.
Total oranges = 9 x 4 = 36
He used 18 oranges.
Number of oranges he had left = 36 − 18 = 18

 **Solution to Question 87**

Sandra sewed 5 buttons on a dress.
She sewed 8 dresses.
Number of buttons used = 8 x 5 = 40

230

She had 12 buttons left.
Total number of buttons she had at first = 40 + 12 = 52

 Solution to Question **88**

Richard bought some bananas for his students.
He gave 5 bananas to each of his 8 students and had 7 bananas left.
Total bananas given to 8 people = 8 x 5 = 40
Therefore number of bananas Richard bought = 40 + 7 = 47

 Solution to Question **89**

Alvin arranged 4 white tables and 5 red tables in a row.
Total tables = 4 + 5 = 9
He arranged 9 such rows
Number of tables altogether = 9 x 9 = 81

 Solution to Question **90**

Jason have 6 boxes of stamps.
There were 12 stamps in each box.
Total stamps = 12 x 6 = 72
Bernice had 32 more stamps than Jason.
Number of stamps Bernice has = 72 + 32 = 104
Number of stamps they have altogether = 104 + 72 = 176

 Solution to Question **91**

There were 9 oranges in Bag A.
There were twice as many oranges in Bag B than in Bag A.
Bag B = 9 x 2 = 18
Number of oranges altogether = 18 + 9 = 27

Solution to Question 92

The big car could seat 8 people and the small car could seat 6 people.
The school hired 4 big cars and 8 small cars.
Total Number of people in 4 big cars = 8 x 4 = 32
Total Number of people in 8 small cars = 8 x 6 = 48
Total Number of people who went for the excursion = 48 + 32 = 80

 ### Solution to Question 93

A packet of apples cost $5 and a packet of mangoes cost $4.
Alicia had just enough money to buy 3 packets of apples and 7 packets of mangoes .
Money she had = 3 x 5 + 4 x 7
= 15 + 28
= $43

 ### Solution to Question 94

There were 9 cars and 10 motorcycles in a car park.
Car wheels = 9 x 4 = 36
Motor cycle wheels = 10 x 2 = 20
Number of more car wheels than the motorcycles wheels
= 36 – 20 = 16

 ### Solution to Question 95

Jason baked some egg tarts and packed them into boxes of 5 each.
He sold 8 such boxes and had 6 boxes left.
Total Number of boxes = 8 + 6 = 14
      a) Number of egg tarts left = 6 x 5 = 30
      b) Number of egg tarts she baked = 14 x 5 = 70

 **Solution to Question 96**

The number of students in a Grade 2 can be divided into 5 groups of 20 students each.
Total number of students = 20 x 5 = 100
15 students are Chinese.
Some students are Americans.
The remaining 19 students are Australians.
Number of American students = 100 − 15 − 19 = 66

 Solution to Question 97

A fruit seller packed the fruits into 5 bags of 8 fruits each.
Total fruits = 8 x 5 = 40
10 of these fruits were apples.
There were 3 more pears than apples = 10 + 3 = 13
The rest of the fruits were oranges.
Number of oranges in the fruit stall = 40 − 10 − 13 = 17

 Solution to Question 98

Total Number of stickers required for all the items
= 1 + 2 + 3 + 4 = 10
Therefore Kevin can get one of each item.
Or he can have 3 stickers and 1 notepad = 2 x 3 + 4 = 10
Or he can have 2 notepads and 2 bookmarks = 2 x 4 + 1 x 2 = 10
Or he can have 2 pencils and 2 erasers = 3 x 2 + 2 x 2 = 10

 Solution to Question 99

Ariana has 5 baskets that hold a total of 15 pears.

233

Each basket has a different number of pears.
Number of pears in each basket = 1, 2, 3, 4, 5

Solution to Question 100

Jane bought 35 oranges.
She put them equally into 5 boxes.
Number of oranges in each box = 35/5 = 7
She gave away 3 boxes to her friends.
Boxes left = 5 − 3 = 2
Number of oranges she had left = 2 x 7 =
14

Solution to Question 101

9 pair of shoes cost $54.
Cost of each pair of shoe = 54/9 = $6
Cost of 2 pairs shoes = 6 x 2 = $12

Solution to Question 102

A mango tree has 8 mangoes.
An orange tree has 10 oranges.
An apple tree has 12 apples.
Two children went picking the fruits and each gets the same number of each kind of fruit.
Number of mangoes each child would get = 8/2 = 4
Number of oranges each child would get = 10/2 = 5
Number of apples each child would get = 12/2 = 6
Number of fruits each child gets in all = 4 + 5 +6 = 15

234

Solution to Question 103

There are some cows and ducks were on a field that
have a total of 76 legs.
There are 10 cows.
Number of legs 10 cows have = 40
Remaining Number of legs = 76 − 40 = 36
Number of ducks = 36/2 = 18
Number of animals altogether = 18 + 10 = 28

Solution to Question 104

I have 16 beautiful plants in my garden.
My garden is divided into 4 equal parts.
II have planted equal number of plants in each part.
Number of plants in each part = 16/4 = 4

Solution to Question 105

Mum calls her 5 children to sit at the table to have their snack.
She has 12 crackers.
She gives each child the same number of crackers = 2 x 5 = 10
Number of crackers each child got = 2
Number of crackers left = 12 − 10 = 2

Solution to Question 106

A mother bought 12 cakes for her children.
She wants to distribute them equally among her 2 children.
Number of cakes each child gets = 12/2 = 6

 **Solut on to Quest on 107**

Kevin reads 2 pages per hour.
Number of hours he will take to complete a 30 page book
= 30/2 = 15 hours

 **Solut on to Quest on 108**

Lucy weighs 32 kg.
Jerry weighs 2 kg lighter than Lucy = 32 − 2 = 30 kg
But Jerry is half as heavy as Sandra.
Sandra weighs = 30 x 2 = 60 kg

 **Solution to Question 109**

A farmer had 260 chicken.
He sold 130 in the morning and 90 of them in the afternoon.
Total number of chicken sold = 130 + 90 = 220
Remaining = 260 − 220 = 40
The rest of the chicken were put equally into 4 packets.
Number of chicken in each packet = 40/4 = 10

 **Solution to Question 110**

Jason bought 6 packets of mangoes.
There were 12 mangoes in each packet.
Number of mangoes in 6 packets = 12 x 6 = 72
He sold the mangoes and put them into bundle of 3.
Number of bundles of mangoes = 72/3 = 24

Solution to Question **111**

Claire baked 5 trays of cupcakes.
There were 8 cupcakes on each tray.
No. of cup cakes in 5 trays = 5 x 8 = 40
She then gave 14 cupcakes to her children and packed the rest equally into 2 bags.
Remaining cup cakes = 40 − 14 = 26
No. of cupcakes in each bag = 26/2 = 13

 ## Solution to Question **112**

Amanda and her sister gathered all 98 of their toys and placed them on the shelves in their bedroom.
Every shelf can carry a maximum of 7 toys
Number of shelves that will be filled = 98/7 = 14

 ## Solution to Question **113**

There was 20 kg of sugar.
Thomas used 14 kg and packed the rest equally into small bags of 2 kg each.
Remaining sugar = 20 − 14 = 6 kg
Number of such bags of sugar she had = 6/2 = 3

 ## Solution to Question **114**

Kelvin had 8 packets of pears.
There were 3 pears in each packet.
Total Number of pears = 8 x 3 = 24

He shared the pears with his 6 friends.
Number of pears each friend got = 24/6 = 4

 Solution to Question **115**

Max and Gavin had 28 oranges.
Max ate 7 oranges and gave away 5 oranges.
= 28 − 7 − 5 = 28 − 12 = 16
They then had the same number of oranges = 16/2 = 8
Number of oranges Gavin had = 8

 Solution to Question **116**

Tim has some marbles.
There are 10 red marbles, 12 green marbles and 2 more orange
marble than green marbles.
Orange marbles = 12 + 2 = 14
Total number of marbles = 10 + 12 + 14 = 36
The marbles were put equally into 4 groups = 36/4 = 9
Number of marbles in each group = 9

 Solution to Question **117**

A cake cost $4.
Lara bought 4 cakes and gave the cashier $60.
Cost of 4 cakes = 4 x 4 = $16
Change that Lara got back = $60 − $16 = $44

 Solution to Question **118**

At a stationery shop, whiteboard markers are sold at 3 for $2.
3 ——— $2

238

? —— $12
$2 + $2 + $2 + $2 + $2 + $2 = $12
Number of markers Daniel can buy with $12 = 3 + 3 + 3 + 3 + 3 + 3 = 3 x 6 = 18

 ## Solution to Question **119**

A pair of shoe cost $10 and a pair of glasses cost $26.
During a sale, the price of all the items fell by $2.
Cost of the pair of shoe = 10 − 2 = $8
Pair of glasses cost = 26 − 2 = $24
Cost of the shoe and the pair of glasses during the sale = 24 + 8 = $32

 ## Solution to Question **120**

Two similar pens cost $8.
Each pen cost = 8/2 = $4
A pen and a pencil cost $6.
$4 + cost of pencil = $6
Cost of pencil = 6 − 4 = $2
Cost of pencil = $2

 ## Solution to Question **121**

A book costs $5, a pencil costs $2 and a pencil box costs
$8. Total cost = 5 + 2 + 8 = $15
Jason want to buy all the items but he needs $3 more.
Amount of money with Jason = 15 − 3 = $12

 ## Solution to Question **122**

Maggie bought 5 books for $10.

Each book costs = 10/5 = $2
Cost of 7 books = 7 x 2 = $14

 Solution to Question **123**

Anna bought a pen that cost $1.50.
She gave the cashier $2.
Remaining change she got = $2 – $1.50 = 50 cents
She was given ten-cent coins and twenty cent coins in change.
2 twenty cent coin and 1 ten cent coins = 50 cents
The maximum number of twenty cent coins that Ann could
have received = 2

 Solution to Question **124**

A pair of socks cost $14 and a pair of shoes cost $18. Doris
bought both items and gave the cashier a $60 note. Total
cost = 14 + 18 = $32
Amount of money received in change = 60 – 32 = $28
She received a ten-dollar note and some two dollar notes in
change.
Total value of $2 notes received = 28 – 10 = $14
Number of two dollar notes she received = 14/2 = 7

 Solution to Question **125**

Belinda spent $45 on 3 watermelons.
Cost of each watermelon = 45/3 = $15
A watermelon cost as much as 3 bananas.
Therefore the cost of one banana = 15/3 =
$5

Solution to Question 126

A pen cost $6.
It cost $3 more than a sharpener.
Sharpener cost = 6 − 3 =$3
Maya had $42.
She bought 2 pens and 2 sharpeners.
Total Cost = 2 x 6 + 2 x 3
= 12 + 6
= $18
Amount of money left with Maya = 42 − 18 = $24

 ## Solution to Question 127

Tim had 6 two-dollar notes, 4 five-dollar notes and 2 one-dollar notes.
Total amount of money he had = 6 x 2 + 4 x 5 + 2 x 1
= 12 + 20 + 2
= $34
He spent $20 on a book and $6 on a magazine.
Total money spent = 20 + 6 = $26
Amount of money left with Tim = 34 − 26 = $8

 ## Solution to Question 128

3 oranges cost as much as 4 apples.
An orange costs $2
Cost of 3 oranges = 2 x 3 = $6
3 oranges cost as much as 3 apples
Cost of an apple = 6/3 = $2

Solution to Question 129

A kiwi fruit and 3 mangoes cost $7.
A mango cost $2
Cost of 3 mangoes = 3 x 2 = $6
Therefore cost of one kiwi fruit = 7 − 6
= $1

Solution to Question 130

Samuel spent $3 on a pair of socks
He spent $14 more on a scarf than on the pair of socks.
Cost of scarf = 14 + 3 = $17
He had $2 left
Amount of money he had at first = 17 + 3 + 2 = $22

Solution to Question 131

Henry earns $10 for every half an hour that he works.
He works from 8.30 am to 12.30 pm
Number of hours Henry worked = 12:30 am − 8:30 pm = 4 hours
Number of half an hours = 4 x 2 = 8
Amount of money he would earn altogether = 8 x 10 = $80

Solution to Question 132

Max gets paid $6 an hour for cleaning his neighbor's yard.
On Saturday, he cleaned from 2:00 pm to 5:00 pm. This means he worked for 3 hours.
Amount of money Max makes = $6 x 3 = $18

Solution to Question **133**

A book cost $6.
Bryan bought 6 such books and gave the cashier a $50 note.
Cost of 6 such books = 6 x 6 = $36
Remaining amount with Bryan = 50 − 36 = $14
The cashier gave her the change in two-dollar notes.
Number of two-dollar notes she got in change = 14/2 = 7

 ## Solution to Question **134**

Kevin bought 8 notebooks.
Each notebook cost $2.
Cost of 8 books = 8 x 2 = $16
He paid for the notebooks with a $50 note
Amount of money he received as change = 50 − 16 = $34

 ## Solution to Question **135**

An magazine costs $4.
Jennifer bought 7 such magazines = 7 x 4 = $28
She had $8 left.
Amount of money she had at first = 28 + 8 = $36

 ## Solution to Question **136**

Paul was queuing for food in a restaurant.
There were 4 children in front of him and 3 children behind him.
Total number of children = 4 + 3 + 1 (Paul himself) = 8
Each of them had $4 each.
Total money with all the children = 8 x 4 = $32

 ### Solution to Question **137**

Dennis bought a watch for $86 and a pen for $12.

Total cost = 86 + 12 = $98

He gave the cashier a $100 note and received his change in twenty cent coins.

Remaining money = 100 − 98 = $2

Number of twenty cent coins in 1 $ = 5

Number of twenty cent coins he received for $2 change = 5 x 2 = 10

 ### Solution to Question **138**

Jessica had $45.

She spent $18 on a toy and the rest of the money on 3 similar T-shirts.

Cost of 3 T- shirts = 45 − 18 = $27

Each T-shirt cost = 27/3 = $9

 ### Solution to Question **139**

Kevin needs to earn 9 dollars.

He earns 1 dollar and 50 cents an hour for working at the library.

He has already worked for 4 hours and earned = 1.5 x 4 = $6

Remaining money required = 9 − 6 = $3

$1.50 + $1.50 = $3.00

Number of hours Kevin needs to work = 1.5 + 1.5 = 2 hours

 ### Solution to Question **140**

James gets 80 cents for his pocket money each Saturday.

On Monday, he bought a pencil for 10 cents.

On Tuesday, he spent 15 cents on candy.

Total money spent = 10 + 15 = 25 cents

Amount of money left with James = 80 − 25 = 55 cents

On Wednesday, he wanted to buy a book that cost 30 cents.
Yes, he had enough money.

Solution to Question 141

An adult ticket to a concert costs $8 and a child's ticket costs $5.
Amount of money 2 adults and a child have to pay for the concert
= 2 x 8 + 1 x 5
= 16 + 5
= $21

Solution to Question 142

Kelvin paid $52 for a pair of shoe.
This was $14 less than what he paid for a jacket.
Therefore jacket's cost = 52 + 14 = $66
Money he paid altogether = 66 + 52 = $118

Solution to Question 143

After buying some stickers for $30.00, Sandra has $7.00 left.
Amount of money Sandra had in the beginning =
$30 + $7 = $37

Solution to Question 144

Carol has $45.00 and Lucy has $8.00. Amount of
more money that Carol has than Lucy = $45 - $8 = $37

Solution to Question 145

Mark receives $15 every month for his pocket money allowance.
He puts $7 out of it into a piggy bank until his piggy bank has $119.
Number of months it took Mark to reach savings of $119 = 7 + 7 + 7 + 7 + 7 + 7 + 7 + 7 + 7 + 7 + 7 + 7 + 7 + 7 + 7 + 7 + 7 = 17 days
Alternatively 119/7 = 17 days

Solution to Question 146

Mr. Carter collected $6 each from his students for their upcoming field trip.
Amount of money collected from all students = $192
Number of students in Mr. Carter's class = 192/6 = 34 students

Solution to Question 147

Carmen gets $20 every week for lunch money.
She sets aside $2 every day.
Number of days it would take her to save up $60 = 60/2 = 30 days

Solution to Question **148**

Emily is selling concert tickets for $4.
She collects a total of $284.
Number of tickets Emily sold = 284/4 = 71 tickets

 Solution to Question **149**

Laura spent $150 at a jewellery store and $50 at a
department store.
Total amount of money spent =150 + 50 = $200
She had $90 left.
Amount of money she had at first = 200 + 90 = $290

 Solution to Question **150**

Melissa bought a shoe that cost $18 and received 2 two-dollar
notes in change.
Two 2$ notes = 2 x 2 = $4
Total amount of money she had at first = 18 + 4 = $22

 Solution to Question **151**

Julia's monthly pocket money allowance is $5.00.
She baby sat her neighbor's kid and earned an extra $1.75.
She bought an ice-cream for $2.25.
On her way home she found $1.00.
Amount of money Julia had = $5 + $1.75 + $2.25 + $1 = $10

247

Solution to Question 152

Jacob got $25.00 from his grandmother.
 He spent $9.00 on buying a book.
Jacob bought a remote control car for $12.00
He bought fizzy soda for $2.50.
Amount of money Jacob has left = $25 - $9 - $12 - $2.50 = $1.50

Solution to Question 153

Louis wants to buy a comic book and it costs $13.00.
He has $6.85.
 Amount of money that Louis needs to buy the comic book
= $13 - 6.85 = $6 and 15 cents

Solution to Question 154

Amy gave the cashier a ten-dollar note.
The cashier returned her 4 two-dollar notes, 3 twenty-cent coins
and 1 ten-cent coin for change.
This means the cashier returned her $8 and 70 cents.
Amount of money Amy spent = $1.30

Solution to Question 155

Keith bought a book for $8.35 and a bag of pens for $3.25.
 He paid with a $20 bill .
Amount of change Keith got back = $20 - $8.35 - $ 3.25 = $8.30

Solution to Question 156

One box of crayons cost $8.
A box of colored pens cost $3 less than a box of crayons = 8 − 3 = $5
Mary used all her money to buy 4 boxes of colored pens = 4 x 5 = $20
Amount of money Mary had at first = $20

Solution to Question 157

Thomas and Mary bought a gift together for their best friend
Lucy..
The gift cost $21.
Thomas paid $3 more than Mary.
Therefore after subtracting $3 from the gift price both
shared equal amount = 21 − 3 = $18
Each shared = 18/2 = $9
Thomas paid = 9 + 3 = $12

Solution to Question 158

There were 6 apple pies and 8 chocolate pies in a box.
An apple pie cost $5 and a chocolate pie cost $4.
Money earned from selling them = 6 x 5 + 8 x 4 = 30 + 32 = $ 62

Solution to Question 159

Mark bought a table and 4 chairs for $200.
He paid $176 for the table.
Therefore cost of 4 chairs = 200 − 176 = $24
The cost of each chair = 24/4 = $6

 **Solution to Question 160**

Mike bought 8 oranges.
Each orange cost $3.
Cost of 8 oranges = 8 x 3 = $24
He paid for the oranges with a $100-note.
Amount of change Mike received = 100 − 24 =
$76

 Solution to Question 161

A pencil cost $3.
A pen cost $4 more than the pencil = 4 + 3 = $7
Alan bought 5 pens.
Cost of pens = 7 x 5 = $35
He gave the cashier four $10-notes= 10 x 4 = $40
Amount of money he received as change = 40 − 35 = $5

 Solution to Question 162

A papaya cost $2.
A watermelon cost $3 more than the papaya = 2 + 3 = $5
Mr. James bought 8 watermelons.
Cost of 8 watermelons = 8 x 5 = $40
He gave the fruit seller a $100 note.
Amount of money he received as change = 100 − 40 = $60

 Solution to Question 163

A diary cost $3.
A dictionary cost $5 more than the diary = 5 + 3 = $8
George bought 4 dictionaries.
Cost of 4 dictionaries = 4 x 8 = $32
He gave the cashier a $50-note.

250

Amount of money he received as change = 50 – 32 = $18

 Solution to Question **164**

A blouse costs $8.
A jacket costs thrice as much as blouse = 3 x 8 = $24
The total cost of the blouse and the jacket = 24 + 8 = $32

 Solution to Question **165**

Jane went shopping with three $50 notes = 50 x 3 = $150
She bought a pen which cost $20 and a dress which cost $40.
Total cost = 20 + 40 = $60
Amount of money she had left = 150 – 60 = $90

 Solution to Question **166**

Kevin went shopping with eighteen $2 notes = 18 x 2 = $36
He spent $29 on a jacket.
Amount of money left with Kevin = 36 – 29 = $7

 Solution to Question **167**

A packet of 3 apples cost $1.50
Cost of 1 apple = 50 cents
A packet of 4 mangoes cost $2.
Cost of 1 mango = 50 cents
Cost of 3 apples and 6 mangoes = $1.50 + $3 = $4 and 50 cents

Solution to Question 168

Daniel wishes to buy 7 mangoes.
Each mango costs $5.
Cost of 7 mangoes = 7 x 5 = $35
He only has $21 with him.
Amount of more money needed = 35 − 21 = $14

Solut on to Quest on 169

A toy car costs $32.
A stuffed toy cost $28 less than the toy car = 32 − 28 = $4
Tony bought 9 stuffed toys.
Cost of 9 stuffed toys = 9 x 4 = $36
He gave the cashier two $20 notes = 2 x 20 = $40
Amount of money Tony received as change = 40 − 36 = $4

Solution to Question 170

A DVD cost $20.
A book cost $15 less than the DVD = 20 − 15 = $5
Jack bought 9 books.
Cost of 9 books = 9 x 5 = $45
He gave the cashier a $50 note.
Amount of money Jack received as change = 50 − 45 = $5

Solution to Question 171

A watch cost $80.
A belt cost $70 less than the watch = 80 − 70 = $10
May bought 5 belts.
Cost of 5 belts = 10 x 5 = $50
She gave the cashier a $100-note.

252

Amount of money Mary received as change = 100 − 50 = $50

 ## Solution to Question 172

Tom has some money that he wanted to use for shopping.
He spent $25 on a jacket and withdrew another $80 from the
ATM. Now, he has a total of $200.
Money Tom had at first − 25 + 80 = 200
Money Tom had at first + 55 = 200
Money Tom had at first = 200 − 55
Money Tom had at first = 145 $

 ## Solution to Question 173

I played soccer at 4:00 o'clock and ate dinner 3 hours later.
Time at which I ate my dinner = 4:00 + 3 hours = 7 o' clock

 ## Solution to Question 174

Kevin's lunch began at 1:05 pm and ended at 2:05 pm.
Number of hours he ate his lunch = 2:05 pm − 1:05 pm = 1 hour

 ## Solution to Question 175

School begins at 9:30 am and gets over at 3:30 pm.
Number of hours you spend at school every day = 3:30 − 9:30 =
6 hours

 ## Solution to Question 176

The first day of a certain month that has 30 days is a Monday.

253

Number of weeks = 30/7 = 4 weeks and 2 days
Number of Mondays this month had = 5

 Solution to Question **177**

A concert started at 8.30 pm and ended at 10.00 pm.
Number of hours the concert ran = 1 hour 30 minutes

 Solution to Question **178**

Samuel is 8 years old this year.
His brother is 3 years older than him.
Present age of his brother = 8 + 3 = 11 years
His brother's age in 5 years' time = 11 + 5 = 16 years

 Solution to Question **179**

Tracy is 36 years old.
Three years ago, her daughter was 9 years old.
Present age of her daughter = 9 + 3 = 12 years
Number of years for her daughter to be 20 = 20 − 12 = 8 years
Age of Tracy when her daughter would be 20 years old = 36 + 8
= 44 years

 Solution to Question **180**

Tom is 12 years old
His mother is 38 years old.
Number of years for Tom's mother to be 60 years old = 60 − 38
= 22 years
Age of Tom when his mother would be 60 years old = 12 + 22
= 34 years

254

Solution to Question **181**

Ted is 18 years old.
His father is 44 years old.
Number of years for Ted's father to be 90 = 90 − 44 = 46 years
Age of Ted when his father would be 90 years old = 18 + 46
= 64 years

 ### Solution to Question **182**

Susan is 47 years old.
Her husband is 8 years older than her = 47 + 8 = 55 years
Their total age in 12 years time = 47 + 55 + 12 + 12
= 126 years

 ### Solution to Question **183**

John is 82 years old.
His grandson is 59 years younger than him = 82 − 59 = 23 years
Their total age in 14 years time will be = 82 + 23 + 14 + 14 = 133
years

 ### Solution to Question **184**

Alfred is 57 years old.
His wife is 9 years younger to him = 57 − 9 = 48 years
Alfred's age 12 years ago = 57 − 12 = 45 years
His wife's age 12 years ago = 48 − 12 = 36 years
Their total age 12 years ago = 36 + 45 = 81 years

 ### Solution to Question **185**

Tom is 16 years younger than his mother.

His father is 7 years older than his mother.
Number of years by which Tom's father is older than him = 16 + 7
= 23 years

Solution to Question 186

Josephine is 24 years older than Ron.
Ron is 67 years younger than Jake.
Number of years by which Jake is older than Josephine = 67 − 24
= 43 years

Solution to Question 187

Kelvin is 4 years old this year.
His brother is thrice as old as him = 4 x 3 = 12 years
His brother's age three years ago = 12 − 3 = 9 years

Solution to Question 188

Amelia bought some flour.
She used 3 kg to bake cupcakes and packed the rest equally into 4
bags of 2 kg each. Total flour in 4 bags = 4 x 2 = 8 kg
Used flour = 3 kg
Total flour = 8 + 3 = 11 kg
Amount of flour Doreen had at first = 11 kg

Solution to Question 189

4 bags of potatoes and 3 bags of rice weigh 45 kg.
Each bag of rice weighs 7 kg.
Weight of 3 bags of rice = 7 x 3 = 21 kg
4 bags of potatoes + 21 = 45 kg
4 bags of potatoes = 45 − 21

4 bags of potatoes = 24 kg
Weight of one bag of potato = 24/4 = 6 kg

 Solution to Question **190**

Weight of the deer is 45 kg less than that of the tiger . Weight of the elephant is 28 kg more than weight of the deer. The tiger is heavier than the elephant.

 Solution to Question **191**

Terry is 18 kg lighter than Jason i.e. Terry < Jason
Rick is 19 kg heavier than Jason i.e. Rick > Jason
Therefore Jason is heavier than Terry.

 Solution to Question **192**

There was 16 kg of flour in bag A.
After some flour was poured from Bag A to Bag B, there was 4 kg more flour in bag A than in bag B.
There was then 9 kg of flour in bag B.
16 – poured flour = 9 + 4
16 – poured flour = 13
Poured flour = 16 – 13 = 3 kg
Amount of flour poured from bag A to bag B = 3 kg

 Solution to Question **193**

A mango is 30 grams heavier than a kiwi fruit.
The mango weighs = 60 grams
Weight of Kiwi fruit = 60 – 30 = 30 grams
Weight of 3 similar kiwi fruit = 30 x 3 = 90 grams

Solution to Question 194

Andy is 24 kg.
Sam is 4 kg heavier than Andy = 24 + 4 = 28
Timothy is 10 kg heavier than Sam = 28 + 10 = 38
The total mass of the 3 boys = 24 + 28 + 38 = 90 kg

Solution to Question 195

A rope is 100 meters long.
It is cut into 2 pieces.
One piece is 32 meters long.
The length of the other piece = 100 − 32 = 68 meters

Solution to Question 196

A box filled with toys weighs 22 kg.
The toys weigh 19 kg.
Weight of the empty box = 22 − 19 = 3 kg
Weight of 5 such empty boxes = 5 x 3 = 15 kg

Solution to Question 197

Joe's weight is 78 kg.
Rick is 19 kg lighter than Joe = 78 − 19 = 59 kg
Leon is 24 kg heavier than Rick = 59 + 24 = 83 kg
Total weight of the three men = 78 + 59 + 83 = 220 kg

Solution to Quest on 198

Weight of a laptop is 2 kg.
A table is 6 times as heavy as a laptop = 2 x 6 = 12 kg
  a) Total weight of the two items = 12 + 2 = 14 kg

b) Number of kg the table is heavier than the laptop = 12 − 2 = 10 kilograms

 Solution to Question **199**

A mug contains 50 milliliters of water.
It contains 45 milliliters less water than a jug. Therefore water in the jug = 50 + 45 = 95 milliliters
The jug contains 60 milliliters more water than a cup. Therefore water in the cup = 95 − 60 = 35 milliliters
The total amount of water in the 3 containers
= 50 + 95 + 35 =180 milliliters

 Solution to Question **200**

Mary has two cows in her house.
Cow A gives 2 liters milk per day and Cow B gives 5 liters milk per day.
Number of liters of milk cow B gives more than cow A = 5 − 2
= 3 liters